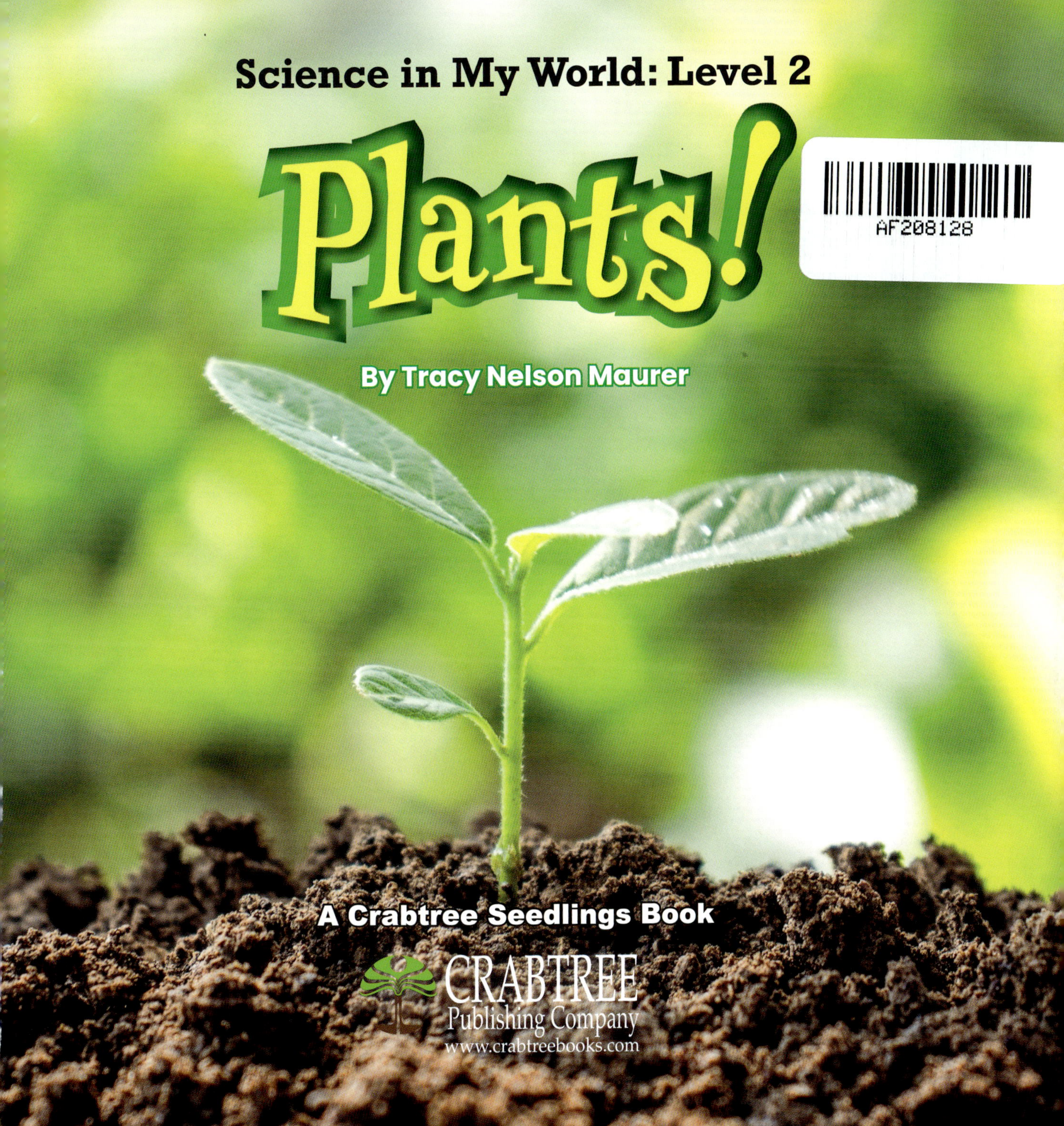

Science in My World: Level 2
Plants!
By Tracy Nelson Maurer
A Crabtree Seedlings Book
CRABTREE
Publishing Company
www.crabtreebooks.com

Table Of Contents

A Planet of Plants

Plants live almost everywhere on Earth. They need sunlight, water, and air to grow. Most need soil for **nutrients**, too.

Fact Digger:
Plants have grown on Earth for more than 500 million years.

Sunflower buds seem to turn to face the Sun through the day.

About 400,000 plant **species** live on Earth. They have adapted, or changed in order to grow well in their **habitats**.

Plants in deserts can store water between rains.

Thick roots reach deeply into the soil in windy places.

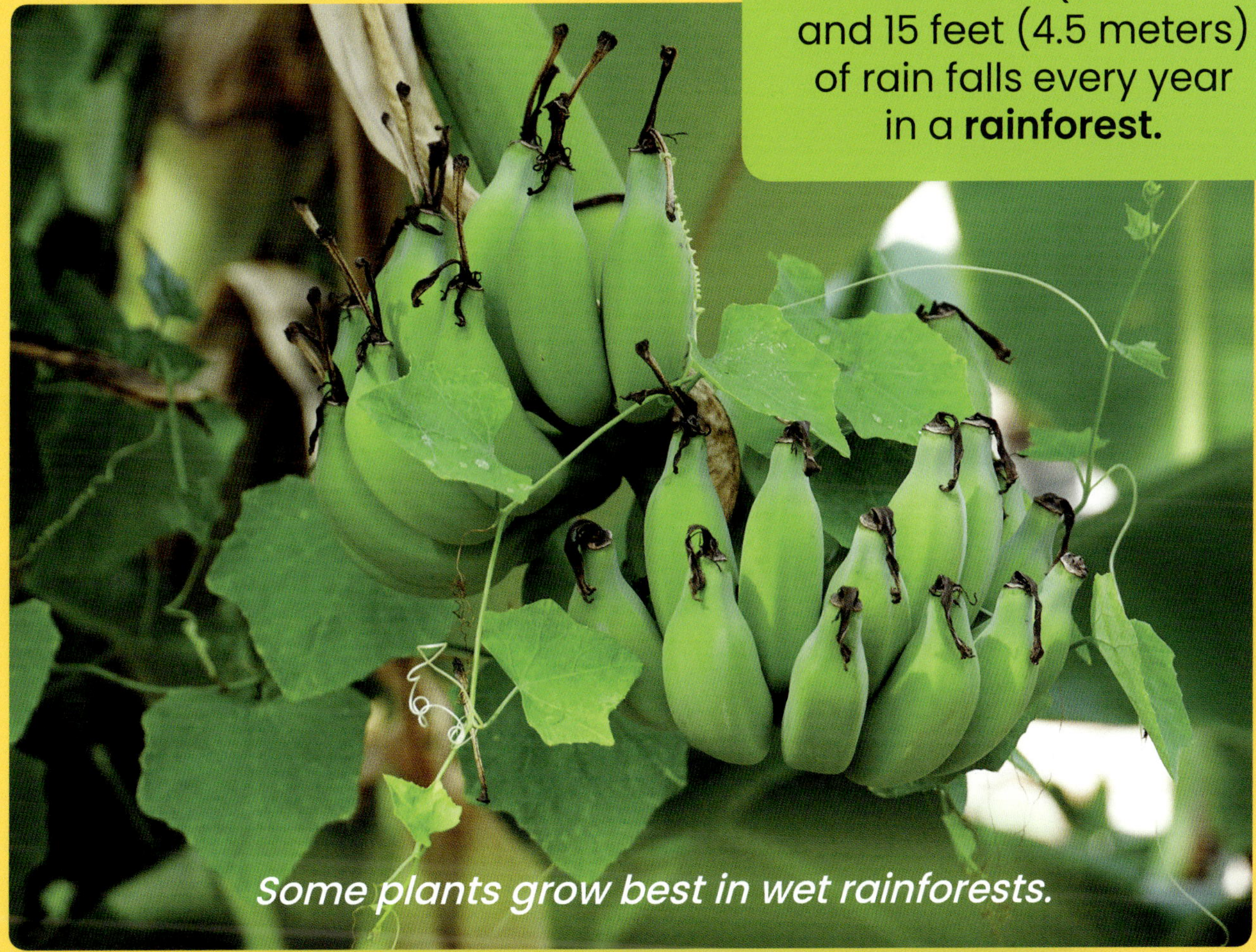

Some plants grow best in wet rainforests.

Plant Parts

Each part of a plant has a job.

- Roots anchor the plant in soil and soak up water and nutrients.

- The stem holds the plant upright and stores water.

- Leaves turn sunlight into food for the plant.

- Flowers or fruits make seeds.

- Seeds grow into baby plants.

leaves
stem
flower
seeds
fruits
roots

Plant parts are many different sizes.

Water-meal fruit is smaller than a grain of salt. The stems of giant sequoia trees can reach more than 274 feet (83.5 meters).

giant sequoia

Food Factories

Most green plants make their own sugary food using **photosynthesis**. Pitcher plants carry out photosynthesis. They also trap insects for dinner.

pitcher plant

During photosynthesis, **chlorophyll** in the leaves captures the Sun's energy.

Tiny holes in the leaves also take in **carbon dioxide** from the air.

Water adds to the mix, too. Leaves release **oxygen** as the plant stores the food it doesn't use.

Fact Digger:
photo means "light" +
synthesis means
"to put together"

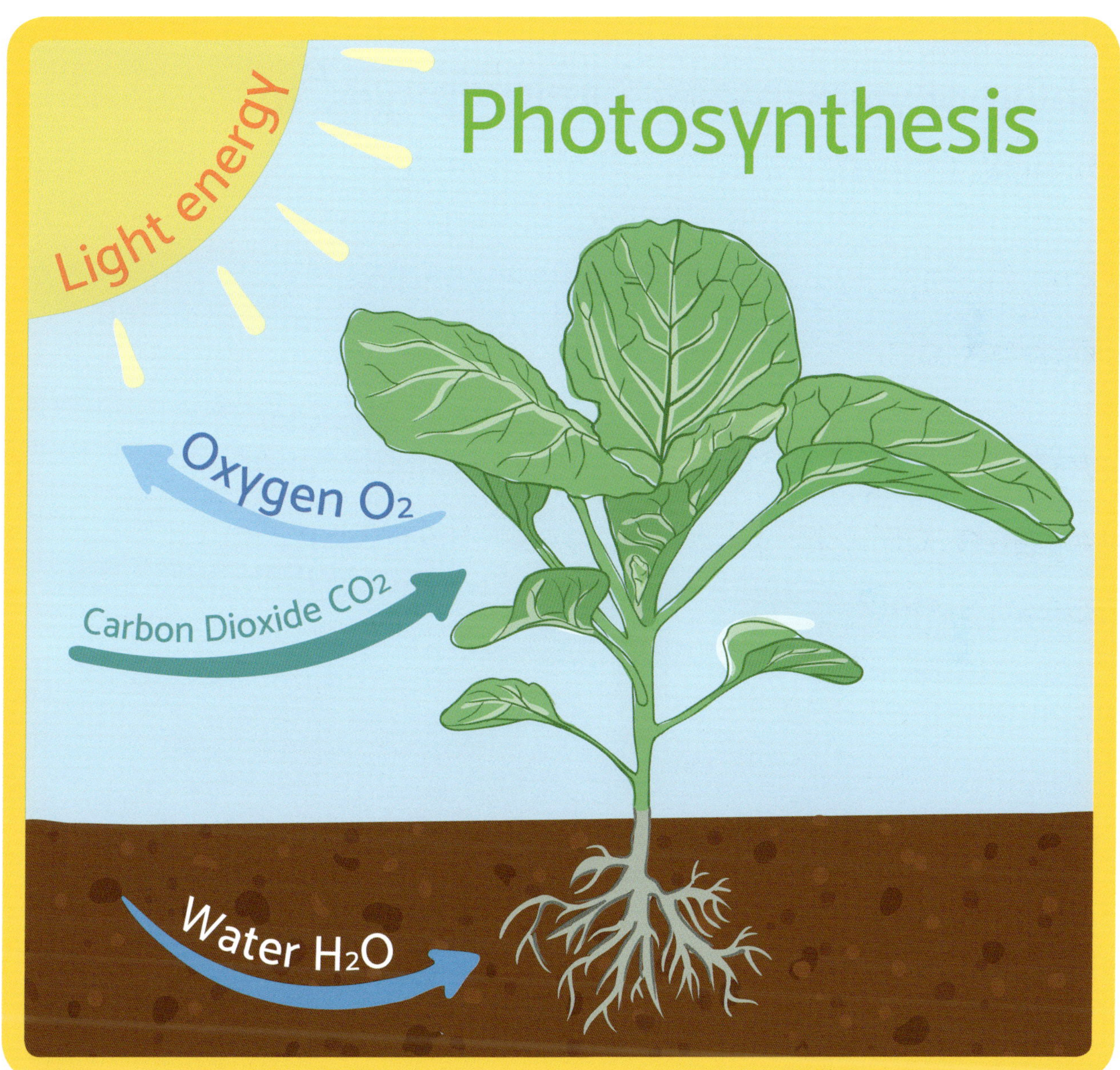

Photosynthesis
Light energy
Oxygen O2
Carbon Dioxide CO2
Water H2O

With less sunlight in autumn, the green chlorophyll in many plants begins to break down.

Instead of green, leaves look
yellow, orange, or red.

Plant Life Cycles

Like all living things, plants have a life cycle:

- A seed sprouts into a baby plant. This seedling is like, but not exactly the same as, its parent.

- The plant makes its own food.

- Adult plants may bloom.

- Flowers may have seeds inside. Some turn into fruits with seeds.

- The fruit ripens and seeds fall to the ground. A seed sprouts into a baby plant.

Many flowering plants use **pollen** to make seeds. But most plants can't make seeds from their own pollen.

Bees, other insects, birds,
and bats can carry pollen
from plant to plant.

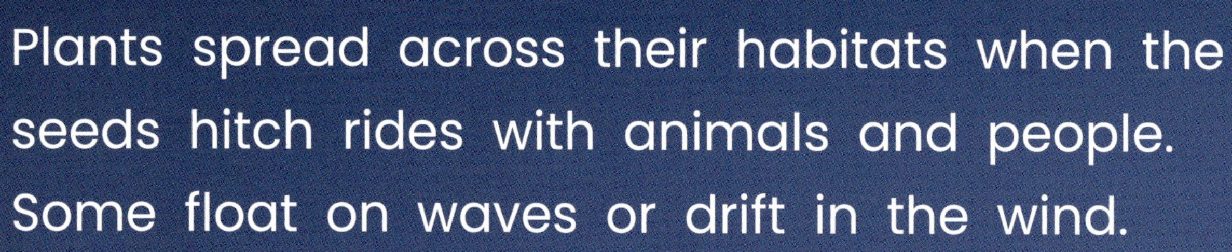

Plants spread across their habitats when the seeds hitch rides with animals and people. Some float on waves or drift in the wind.

Dandelion seeds sail far away from the parent plant.

Some plants grow from underground stems instead of seeds.

Baby ferns come from spores. Spores need water and a mix of other spores to become an adult.

Rhizomes are thick, underground plant stems.

Some long stems can send down roots for a new plant.

Plants protect themselves to live throughout their life cycles. Some taste bad. Some are **toxic**. Pointy parts can also keep away hungry mouths.

The cactus protects itself with sharp needles.

People Need Plants

People and animals need plants for oxygen and food. People also use plants for paper, fabric, lumber, and medicines.

Glossary

adapted (uh-DAPT-tid): Changed or adjusted to new conditions

carbon dioxide (KAHR-buhn dye-AHK-side): A gas made of carbon and oxygen that plants use in photosynthesis and people and animals breathe out

chlorophyll (KLOR-uh-fil): The green substance plants use to make food

habitats (HAB-i-tats): Places where plants usually grow

nutrients (NOO-tree-uhnts): Materials in soil that help feed a plant and keep it healthy

oxygen (AHK-si-juhn): A gas in the air and water that humans and animals need to breathe

photosynthesis (foh-to-SIN-thi-sis): A process that plants use to make food from sunlight, water, and carbon dioxide

pollen (PAH-luhn): Tiny grains that flowering plants use to make seeds for new plants

rainforest (REYN FOR-ist): A forest that receives a lot of rain and that has very tall trees

species (SPEE-seez): Groups set up by scientists to show which plants and animals are alike and can mate with each other

toxic (TAHK-sik): Harmful or poisonous

Index

School-to-Home Support for Caregivers and Teachers

This book helps children grow by letting them practice reading. Here are a few guiding questions to help the reader build his or her comprehension skills. Possible answers appear here in red.

Before Reading

- **What do I think this book is about?** *I think this book is about sunflowers. I think this book is about plants and how they use sunlight.*

- **What do I want to learn about this topic?** *I want to learn the different parts of a plant. I want to learn about photosynthesis.*

During Reading

- **I wonder why...** *I wonder why there are so many different plant species on Earth. I wonder why bees and other animals are needed to spread pollen to different plants.*

- **What have I learned so far?** *I have learned there are tiny holes in leaves that take in carbon dioxide from the air. I have learned that plant leaves release oxygen as the plant stores the food it doesn't use.*

After Reading

- **What details did I learn about this topic?** *I have learned that roots anchor the plant in soil, and soak up water and nutrients. I have learned that leaves need sunlight to make food for the plant.*

- **Read the book again and look for the glossary words.** *I see the word **nutrients** on page 4, and the word **oxygen** on page 14. The other glossary words are found on page 30.*

Library and Archives Canada Cataloguing in Publication

Title: Plants! / by Tracy Nelson Maurer.
Names: Maurer, Tracy Nelson, 1965- author.
Description: Series statement: Science in my world: level 2 |
    "A Crabtree seedlings book". | Includes index.
Identifiers: Canadiana (print) 20210206462 |
    Canadiana (ebook) 20210206470 |
    ISBN 9781039600409 (hardcover) |
    ISBN 9781039600478 (softcover) |
    ISBN 9781039600546 (HTML) |
    ISBN 9781039600614 (EPUB) |
    ISBN 9781039600683 (read-along ebook)
Subjects: LCSH: Plants—Juvenile literature.
Classification: LCC QK49 .M38 2022 | DDC j580—dc23

Library of Congress Cataloging-in-Publication Data

Available at the Library of Congress

Crabtree Publishing Company

www.crabtreebooks.com   1–800–387–7650

Print book version produced jointly with Blue Door Education in 2022

Written by Tracy Nelson Maurer

Print coordinator: Katherine Berti

Printed in the U.S.A./062021/CG20210401

Content produced and published by Blue Door Publishing LLC dba Blue Door Education, Melbourne Beach FL USA. Copyright Blue Door Publishing LLC. All rights reserved. No part of this book may be reproduced or utilized in any form or by any means, electronic or mechanical including photocopying, recording, or by any information storage and retrieval system without permission in writing from the publisher.

Photo Credits: www.shutterstock.com. Cover © Cindy Gokey; Fact Digger icon © Sky vectors; title page ©shutterstock.com/MEE KO DONG. Pages 2-3 ©shutterstock.com/Zadorozhnyi Viktor. pages 4-5 © Cindy Gokey; page 6 cactus © Ilyshev Dmitry, page 6-7 grasses ©hayaka tung. page 7 ©shutterstock.com/Athawit Ketsak. page 9 ©Ardely; page 10 ©Magnetic Mcc, page 11 ©Aleksei Potov; page 12-13 ©chockdee Romkaew; page 15 © Merkushev Vasiliy; page 16 and 17 ©Smit; page 19 ©Kazakova Maryia; page 20 ©Eduardo Dzophoto; page 21 ©shutterstock.com/Ondrej Prosicky. page 22-23 ©Alexey Torbeev; Page 24 closeup ©Phattaraphum, fern ©Thanaporn Pinpart, page 2-25 rhizome © Nadzeya Pakhomava, strawberry plant ©Elena Masiutkina; page 26-27 ©Roger M Buss; page 28-29 ©Pinkyone; page 31 ©shutterstock.com/Artur_eM All images from Shutterstock.com

Published in the United States
Crabtree Publishing
347 Fifth Ave.
Suite 1402-145
New York, NY 10016

Published in Canada
Crabtree Publishing
616 Welland Ave.
St. Catharines, Ontario
L2M 5V6